The Messerschmitt Foundation's Messerschmitt Me 262 A-1a (D-IMTT) on approach to land at Manching airport.

Me 262 and its Variants

Manfred Griehl

The first serious discussions about installing turbojet engines in fighter aircraft took place in early 1938.

The first project work began in October 1938 following the definition of basic requirements. Officially, work on a jet-propelled fighter aircraft began on 01/04/1939 with the issuing of project guidelines for a "high speed fighter aircraft".

The first project proposal for the "P 65 Pursuit Fighter with BMW P 3304 Power Plants" was sent to the RLM's Technical Office (Technische Amt) at the beginning of May. The aircraft was a cantilever low-wing monoplane of all metal construction with a fully retractable undercarriage. The envisaged power plants were two BMW 3302 turbojets, with which the aircraft was expected to have a cruising speed of 900 kilometers per hour at 100% thrust. It was believed that the two turbojets could be accommodated in the wing center-section of the P 65, thus reducing drag.

During an installation meeting between Messerschmitt and BMW on 01/12/1939, the possible use of the BMW 3304 engine, mounted beneath the wings of the low-wing monoplane, was also discussed. Meanwhile the first visual and cockpit mockup of the Me P 65 was completed. It was inspected by representatives of the Technical Office (LC 2) and the Rechlin Proving Station on 19/01/1940.

There was much work to be completed at the beginning of 1940: construction of full mockups,

The Me 262 V3 (PC + UC) during ground trials.v

This rare photo depicts the Me 262 V1 with Jumo 210 piston engine (left in photo) prior to the installation of jet engines. In the foreground is a Bf 108 (TJ + AX).

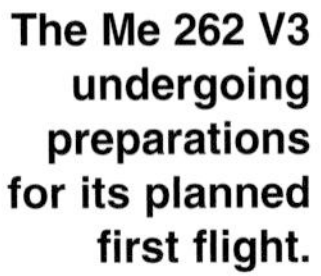

The Me 262 V3 undergoing preparations for its planned first flight.

Fritz Wendel beside the Me 262 V3 after its successful first flight.

aerodynamic experiments, wind tunnel testing, and construction of the first prototype. A Jumo 210 G was to be installed in the weapons compartment of the first prototype, as the BMW engines still appeared to be nowhere near the desired production readiness.

At a meeting on 01/03/1940 it was decided to equip the Me P 65 V1 with an ejection seat, air brakes and a braking parachute. It was also to be fitted with a pressurized cockpit and protected fuel tanks. An initial order for three prototypes was placed, all with BMW turbojet engines.

The equipment list for the P 65 was revised several times by May, and on 15/05/1940 a second project proposal was submitted to the RLM. As it had turned out that the jet engines could not be housed within the wing structure (on account of the spar), the company now proposed underslung engines and a swept wing. The reason for the latter was the change in the center of gravity caused by the increased weight of the two jet engines. On 08/04/1941 consideration was given to equipping the Me P 65 with two rocket engines made by the Walter company, as it was expected that these would become reliable much sooner than the BMW P 3302 or P 3304 engines previously considered.

Initial flight trials began on 18/04/1941, although the sole power plant was the Jumo 210 G piston engine. Although there was still no information as to how the airframe would behave in the higher speed range, on 25/07/1941 the LC 2 (T.St.Nr. 6303/41) ordered the construction of five prototypes and twenty pre-production aircraft. Soon afterward, on 04/08/1941, Beauvais and Bader of the Rechlin Proving Station test-flew the first prototype of the Me 262 – still without turbojet engines. Subsequent to this a brief specification was written for a "reconnaissance aircraft", which was to come off the production line alongside the "pursuit fighter". An initial draft was presented to the Technical Office at the end of October 1941.

At the beginning of 1942 a mockup of the projected reconnaissance version was under construction, along with the Me 262 V2 and V3 prototypes.

On 25/03/1942 the Me 262 V1 barely avoided a crash, thanks to its Jumo 210 G piston engine. Fritz Wendel took off in PC + UA at 19:29 and just five minutes later the prototype was back on the ground.

Soon after takeoff, fuel pressure dropped sharply and both BMW 3302 jet engines quit. The following is from flight report No. 692/32, written by company test pilot Wendel on 30/03/1942: "For lack of time I was unable to pay any particular attention to the aircraft's characteristics. I only noted the high aileron forces, which felt even higher than before. The sink rate while on approach to land is very high. Even though I maintained an approach speed of 240 kph, the wing slats were open all the time and the aircraft was difficult to control. Landing was normal; after 2/3 of the landing run, both torque links on the undercarriage legs broke for reasons that are as yet unknown.

The overall impression was less than satisfactory. It can now already be said that average pilots cannot master this aircraft with such a high wing loading."

Not until 17/07/1942 was Fritz Wendel able to complete the first successful, but in particular longer, jet-powered flight in the Me 262 V3 (PC + UC) at Leipheim. On 11/08/1942, however, after just six flights by Flugkapitän Wendel, Fliegerstabsing. Heinrich Beauvais was unable to get the V3 to lift off at Rechlin, resulting in a crash.

As the BMW turbojets were considered far from operationally mature, the Me 262 V2 also had to be fitted with a pair of experimental Jumo engines. Wendel carried out the first flight in the machine in Lechfeld on 01/10/1942, after which Beauvais took over test flying in Rechlin.

In the meantime practical trials had resulted in numerous improvements in the aircraft's general equipment. Significant steps had also resulted in changes in the appearance of the Me 262 prototypes. On 02/10/1942 the RLM ordered that, in addition to the Me 262 V1 to V5, the next five prototypes and 30 pre-production machines should also be built. The first prototype with a nosewheel undercarriage was supposed to be ready for trials by the end of 1943. Only a few days later the Messerschmitt company criticized this deadline as unfeasible from a production point of view. Nevertheless, new plans for the prototypes and pre-production machines were issued on 02/12/1942. The Technische Amt (GLJC-E2) directed that the first aircraft with the provisional nosewheel undercarriage should be completed as quickly as possible. As well, instructions arrived from Berlin to significantly increase the size of the planned pre-production series. As of the beginning of 1944, 20 Me 262s were to be completed each month with a significantly improved armament. Instead of three MG 151s, the aircraft would have four 30-mm weapons or one MK 103 and two MG 151/20s.

These steps were taken as penetrations into German airspace by Allied long-range bombers were becoming ever more frequent.

A significant change in the Me 262's aerodynamics resulted from the addition by Messerschmitt of a filet to achieve a straight leading edge, to avoid flow separation on the inner wing. A Me 262 fuselage purposely built for free-fall experiments was supposed to be dropped from the Me 323 S9 over Lake Müritz near Rechlin, however this proved impossible when the lake froze over. For this reason the test was repeated over the Chiemsee on 11/02/1943. With the assistance of Flak Brigade 4 from Munich, which provided the measuring team, the Me 262 fuselage was released at an altitude of 6000 meters. Measurements revealed that the fuselage reached a maximum speed of 870 kph at an altitude of about 2000 meters. This dropped off to about 800 kph by the time it hit the surface of the lake. As the parachute was destroyed when it deployed, the experiment had to be considered a failure and was supposed to be repeated over Lake Constance on 23/10/1943. This time a glider-tug combination was used, consisting of a He 111 Z and a Me 321 glider with underslung Me 262 fuselage. This test was also a failure, however, as the parachute failed to fully deploy.

Me 262 Prototypes and Pre-Production Aircraft

Type	Code	WorkNr.	First Flight	Remarks
Me 262 V1	PC+UA	262 000001	25.3.42	Crashed: 7.6.44
Me 262 V2	PC+UB	262 000002	1.10.42	Crashed: 18.4.43
Me 262 V3	PC+UC	262 000003	18.7.42	Air raid: 12.9.44
Me 262 V4	PC+UD	262 000004	15.5.43	Crash landing. 25.7.43
Me 262 V5	PC+UE	262 000005	16.6.43	Crashed: 1.2.44
Me 262 V6	VI+AA	130001	17.10.43	Crashed: 9.3.44
Me 262 V7	VI+AB	130002	20.12.43	Crashed: 19.5.44
Me 262 V8	VI+AC	130003	18.3.44	Crashedlandg. Okt. 44
Me 262 V9	VI+AD	130004	19.1.44	Present April 44
Me 262 V10	VI+AE	130005	15.4.44	Present April 45
Me 262 S1	VI+AF	130006	19.4.44	Damaged: April 44
Me 262 S2	VI+AG	130007	28.3.44	Damaged: Juni 44
Me 262 S3	VI+AH	130008	16.4.44	Damaged: April 44
Me 262 S4	VI+AI	130009	Mai 44	Damaged: Mai 44
Me 262 S5	VI+AJ	130010	April 44	Crashed: 8.10.44
Me 262 S6	VI+AK	130011	April 44	Crashed: 18.7.44
Me 262 S7	VI+AL	130012	Mai 44	Unknown
Me 262 S8	VI+AM	130013	Mai 44	Crashed: 19.7.44
Me 262 S9	VI+AN	130014	Mai 44	Unknown
Me 262 S10	VI+AO	130015	Mai 44	Present April 45
Me 262 S11	VI+AP	130016	Mai 44	Unknown
Me 262 S12	VI+AQ	130017	Mai 44	Crashed: Okt. 44
Me 262 S13	VI+AR	130018	Juni 44	Crashed: Okt. 44
Me 262 S14	VI+AS	130019	Juni 44	Present Juli 44
Me 262 S15	VI+AT	130020	Juni 44	Shelled.: Jan. 45

The Me 262 V2 shortly before the start of factory testing with Jumo 004 power plants. The aircraft crashed on 18/04/1943, killing company test pilot Ostertag.

Below: Takeoff by the Me 262 V3. After 149 flights, on 12/09/1944 the prototype was badly damaged in an Allied air raid.

However, back to the Me 262 V1. In March 1943 the first prototype was finally converted to Jumo 004 turbojets. As the pressure suits on hand in Rechlin were unsuitable for the cramped cockpit of the V1, the aircraft was made provisionally pressure-tight so that high-altitude testing with two jet engines could begin as soon as possible. The first flight by the converted Me 262 V1, now armed with three MG 151 cannon, took place on 19/07/1943.

Several months earlier Hptm. Späte (EK 16) had test-flown one of the early prototypes, after which he remarked, probably prematurely, that: "...the machine is already in a condition in which it can be deployed immediately!" There was still plenty to do, however, before there could be talk of an operationally-ready Me 262 A-1a. On 18/04/1943 Messerschmitt test pilot Ostertag crashed near Lechfeld due to an inadvertent tailplane setting and was killed.

The Me 262 V4 began flight trials on 15/05/1943. At the same time, Generalmajor Galland, Hptm. Behrens and Fliegerstabsing. Beauvais test flew the third and fourth prototypes.

After this, the word quickly spread that this new fighter possessed a performance that was superior to all aircraft in use by the Allies, an advantage that must not be squandered. As a result of Galland's evaluation, at a meeting held by the state secretary on 25/05/1943 the order was given to assign priority to the production of the Me 262. One-hundred examples of the new fighter were to be delivered by 31/12/1943. The fighter's armament was provisionally defined as three MG 151/20 cannon with 320 rounds per gun. At the same time it was directed that priority be given to the creation of a new nose accommodating six MK 108 cannon. The KdE, Oberst Petersen, was named "Commissar for the Me 262" in order to accelerate the "startup of the type".

In the summer of 1943 the Me 262's undercarriage was strengthened substantially and the brake system improved. As well, on 07/06/1943 the Me 262 V5 finally made its maiden flight, although it was still fitted with a nosewheel undercarriage taken from the Me 309. Eleven days later the aircraft was successfully test-flown in Rechlin.

The damaged Me 262 V4 in a hangar in Lechfeld. It was subsequently DY on 01/02/1944.

Fritz Wendel took the Me 262 V3 into the air for the first time on the morning of 18 July 1943. This photo was taken during taxi distance measurements by company Pilot Ostertag in Leipheim. Parked in the background are Me 321 gliders.

Side view of the Me 262 V3 (PC+UC), seen here taxiing our for another test flight. The design of the tailwheel undercarriage proved disadvantageous and the switch was made to a nosewheel undercarriage on subsequent prototypes.

On 25/07/1943 an Me 262 – the fourth prototype – was demonstrated to the Reichsmarschall for the first time in Rechlin-Lärz by Messerschmitt test pilot Lindner. On its way back to Lechfeld the aircraft made a stopover in Schkeuditz. Although the runway had been lengthened, when the V4 attempted to take off it failed to get airborne in time, resulting in a serious crash. Several days later, on 04/08/1943, the V5 was seriously damaged at Lechfeld after the nosewheel tire burst. It subsequently underwent repairs until November 1943.

The first significantly improved prototype, the Me 262 V6, began flight testing on 17/10/1943. It was powered by two Jumo 004 B-1 engines. Several days earlier, on 3 October, Lindner had achieved a speed of 950 kph at an altitude of 5000 m while flying the Me 262 V3. Following these positive results, on 02/11/1943 Reichsmarschall Göring visited the Messerschmitt factory in Regensburg and the flight testing establishment in Lechfeld, where the Me 262 V6 gave a flying demonstration. On 26/11/1943, at the instigation of the RLM, the prototype was demonstrated to Hitler in Insterburg. The machine was subsequently test-flown by Major Maier (JG 2 "Richthofen") and Hptm. Thierfelder (EK 262) on 21/12/1943.

About 24 hours earlier the Me 262 V7, powered by two Jumo 004 B-1 power plants, had also begun flight testing. The V9 followed on 19/01/1944. There was another accident on 01/02/1944 when Hptm. Thierfelder crashed while taking off from Lechfeld.

The Allies were not unaware of Messerschmitt's activities and especially the acceptance and flight testing operation in Lechfeld. After careful preparations, on 25/02/1944 a force of 268 B-17s of the 1st Bomber Division attacked the Augsburg-Haunstetten-Lechfeld area. The Me 262 V8 and V10 were damaged. The Me 262 V6 was also lost in Lechfeld on 08/03/1944 while being flown by Kurt Schmidt. Karl Mayer, a control group leader attached to Messerschmitt, witnessed the crash:

"I was at the Me 163 run-up pad and watched as an Me 262 taxied to the runway and then took off. The takeoff, in the direction of Augsburg, was normal and I quickly lost sight of the machine.

When the accident took place I was still at the run-up pad and was looking in the direction of Hangar 3, as I was expecting several people from there.

The Me 262 V6 (VI+AA) was the first prototype with a fully functioning tricycle undercarriage..

The Me 262 V7 (VI+AB) during flight trials.

The Me 262 V9 (VI+AD) was tested with a streamlined canopy in October 1944.

The Me 262 V9 and V10 on Lechfeld's scrap heap in the summer of 1944.

A high-pitched whine caught my attention, and I immediately observed an aircraft fly past from the right (west) at, in my estimation, at least 1000 kph. I believe the pilot intended to descend slightly and then pull up over the airfield. I estimate that the machine was at a height of about 300 meters and its dive angle was approximately five degrees. I was just about to draw the attention of my colleague Fleck, who was standing beside me, to the aircraft's amazing speed, and called out to him, "Look there...", when the machine suddenly deviated from its previous inclination and in seconds disappeared behind the hangar. At once debris flew over the hangar. The whole thing happened so quickly that my colleague Fleck only saw flying debris from the crash even though he reacted immediately to my call..."

If that wasn't bad enough, ten days later there was another major air raid which, in addition to Landsberg/Lech, Oberpfaffenhofen and Memmingen, also struck Lechfeld. On 18/03/1944 a delayed-action bomb damaged the left wing and engine of the Me 262 V10. The Me 262 V1 and V3 sustained minor damage from bomb fragments. The Me 262 V7, which was then parked without engines, escaped damage. The pre-production aircraft with the serial numbers 130006 to 130008 were not as lucky, receiving varying degrees of damage from fragmentation bombs.

The loss of trained personnel and equipment caused by the air raids was also a serious blow. A direct hit on a slit trench killed six instrument installers. The attack also struck 1./KG 51, 1./KG 100 and IV./KG 40, all stationed at Lechfeld. After the attack more blast pens were built and the relatively few jet fighters were dispersed over as wide an area as possible. Precision raids on production facilities delivered several more serious blows to the Me 262 program in 1944. Deliveries of the Me 262 A-1a to operational units were often halting. After effects of the war prevented any Me 262s from being delivered in May 1944, in June KG 51 "Edelweiss" received its first five Me 262s. The so-called "Blitz Bomber" now enjoyed the highest priority. In July another 50 aircraft were delivered, along with six converted machines and 14 prototypes.

Of nine Me 262 A-1s delivered in August 1944, five were assigned to the General der Jagdflieger's Test Commando (Fighter) and three to KG 51, while the last Me 262 was assigned to Fliegerüberführungsgeschwader 1 (FLÜG 1, or Aircraft Ferry Wing 1) for training purposes.

According to the records of the Quartermaster-General Dept. 6 IIIC, in September 1944 17 Me 262 A-1s were delivered, along with four two-seaters (Me 262 B-1a): two to KG 51 "Edelweiss", one to KG 54 (J) and the fourth to the Test Commando (Fighter). Despite overcrowding of the airfield, Me 262 test flying had to be completed as quickly as possible, even though air raid alarms significantly hindered flying operations in Lechfeld from the late summer of 1944.

Messerschmitt company report No. 38, regarding the overall Me 262 situation and dated 10/08/1944, has survived. It shows that

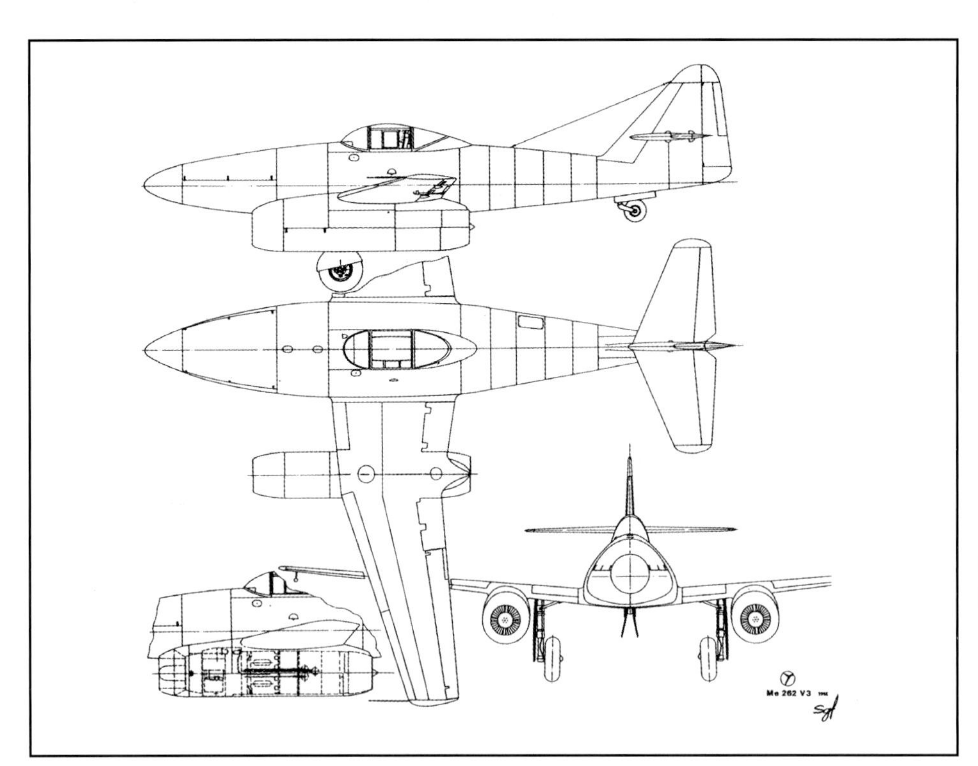

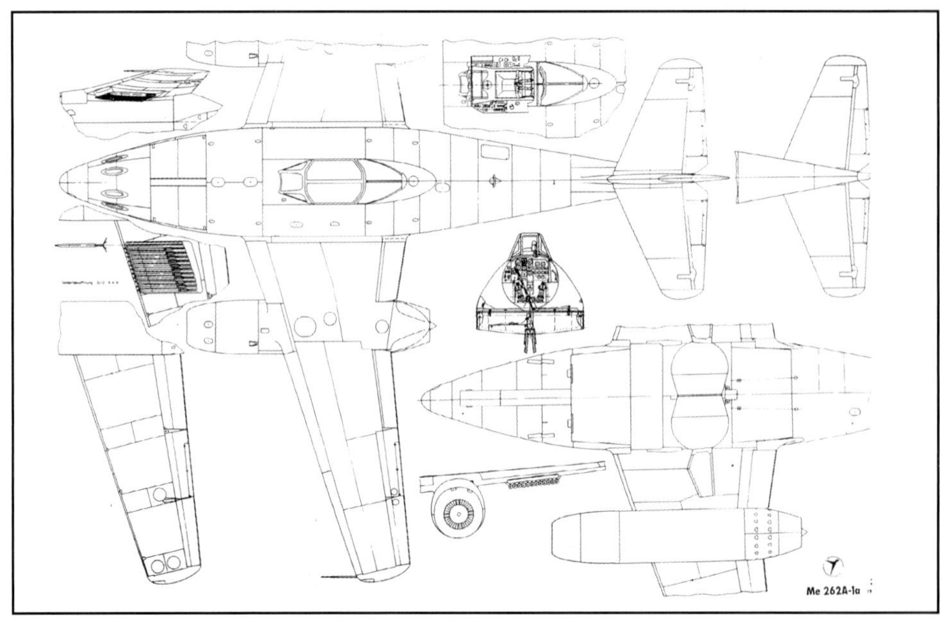

A comparison of these drawings the different wing geometries of the Me 262 V3 and the Me 262 A-1a production aircraft.

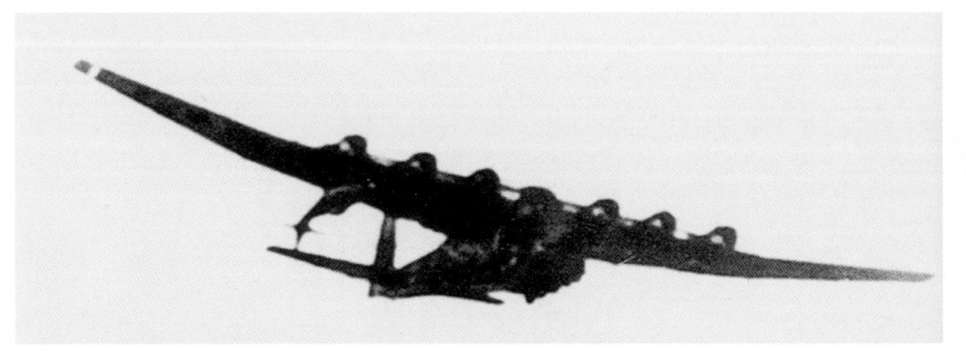

A Me 262 fuselage mounted beneath the wing of a Me 323 "Gigant". The transport carried the fuselage to altitude for drop tests.

The Me 262 V6 (VI+AA) is demonstrated to Göring at Lechfeld on 02/11/1943.

the ambitious plans to complete 100 Me 262s by the end of 1943 remained an illusion: altogether the company had by then completed ten prototypes and 112 production aircraft. Of the prototypes, in August 1944 only two were still engaged in the Messerschmitt company's test program. One Me 262 had meanwhile been released to the air force as a training machine. The Me 262 V1, V2 and V4 to V7 had been lost to the test program or destroyed in crashes. Of the 112 Me 262s, just 80 were available at the beginning of August 1944:

KG 51		33
Test Commando		15
Proving Center	14	
Messerschmitt trials	7	
Converted aircraft:		4
Jungk company		1
Blohm & Voss company	10	
		—
		80

Allied air raids and accidents had reduced the number of series production aircraft to 32. Nine Me 262s were lost n a raid on Leipheim on 24/04/1944. Twelve more were destroyed in attacks on Lechfeld and Leipheim on 19/07/1944. Defective fuel systems resulted in the crashes of three aircraft during delivery, five more failed to reach their destinations due to engine trouble. Two of the pilots lost their lives. A third pilot (KG 51) was killed when his "Blitz Bomber" struck the water and crashed.

Apart from the Me 262 V9 and V10, between 28/09/1944 and 18/10/1944 the only aircraft in Lechfeld were those with the Werknummer 130015 (Baldrian System), 130167 (stability/braking system), 170056 (stability), 170079 (brakes), 130186 (the later Me 262 C-1a) and 170303 (dropped weapons). Until the end of the war, special urgency was given to development of the "Heimatschützer" and the "Lotfe Bomber" (Me 262 A-2a/U2).

Jet flying in Lechfeld temporarily came to an end in April 1945. On 23/04/1945 III.(Erg.)/JG 2 moved to Munich-Riem. Four days later troops of the 4th Infantry and 12th Armored Divisions occupied the airfield. The 308th Bomb Group of the American 8th Air Force and a P-47 fighter unit subsequently moved into Lechfeld. They shared the field with "Watson's Whizzers", which restored captured Me 262s to flying condition. Today Lechfeld is home to the Luftwaffe's Jagdbombergeschwader 32.

Below: The Me 262 S-3 (WerkNr. 130 008) after its accident on 16/04/1944. It rejoined the test program following repairs.

A pair of Me 262s about to take off from Lechfeld.

Center: An operational aircraft of III./ JG 7 finished in the then standard fighter camouflage scheme. The "S" in the center of the fuselage cross indicates that the aircraft was being used as a trainer.

Messerschmitt Me 262 WerkNr. 110 604. The aircraft came from the Schwäbisch-Hall production Line.

The First Aerial Victory of a Jet Fighter in the History of Aviation was Won with this Me 262 A-1a!

The first aerial victory by a jet fighter in the history of aviation was achieved in this Me 262 A-1a.

Flying Me 262 A-1a "White 4", WerkNr. 130017, Lt. Alfred "Bubi" Schreiber scored the first aerial victory by a jet fighter on 26 July 1944. His victim was a Mosquito. Note the white victory bar above the Werknummer. The fuselage band was probably yellow. On 26/11/1944 Schreiber was killed in this machine when it crashed while on approach to land at Lechfeld. Fw. Josef Pellmaier of Erprobungskommando 262 had been tasked with towing the Me 262 from the end of the runway with his Büssing prime mover. He placed the vehicles in 1st gear, hopped out and took this photo of the "historic bird".

In the cockpit of the Me 262 is Pellmaier's co-driver. Photographing the "turbos" was strictly forbidden.

The successful jet pilot Lt. Schreiber with bouquet in front of his aircraft. Note the powder stains around the gun port of the lower cannon.

The successful jet pilot, Lt. Schreiber, stands before his plane holding a bouquet. Note the traces of muzzle flash by the lower gun. Weber lost his life with JG 7 on March 21, 1945, as told in Manfred Boehme's book "Jagdgeschwader 7".

This Messerschmitt Me 262 A-1a, WerkNr. 130017, scored the first aerial victory in history by a jet fighter. Pilot was Lt. Schreiber.

Aftermath of a crash-landing by the third Me 262 A-1a production aircraft. The aircraft was used mainly to test the nosewheel undercarriage.

The nose gear proved an ongoing weak point and caused numerous accidents. For this reason the aircraft could only be towed using a special fork with cables attached to the main undercarriage.

A fuel truck tows one of the two Me 262 A-2a/U3s (planned series designation A-3) and a Me 262 A-2a (WerkNr. 130170)..

Changing an engine of the Me 262 A-2a/U3 V55. Just two examples of this variant were built for testing and it never entered quantity production.

Three views of the Me 262 V167 during testing activity.

This Me 262 A-1a/U4 (WerkNr. 170083) was captured by the Americans in Lechfeld. They subsequently named the aircraft "Wilma Jeanne".Jeanne" erhielt.

Messerschmitt Me 262 "Pulkzerstörer"

The initial step toward the first Me 262 Pulkzerstörer (literally "formation destroyer") was a 1:1 mockup. This was followed by two flying prototypes converted from standard A-1a production aircraft.

The development was the result of a directive from armaments minister Speer dated 05/01/1945, which took into account Hitler's order of 04/11/1944 that "heavy long-range cannon" should be installed in aircraft. In addition to the Ju 88 and Ju 388, the Do 335 and Me 262 in particular were supposed to be fitted with super-heavy weapons. The two choices were the Rheinmetall-Borsig BK 5 and the Mauser 214A. Because of its low rate of fire, just 45 rounds per minute compared to 150 rounds per minute, the BK 5 had no real chance. Attempts were also made to install the MK 112 and MK 412 in the 8-262, however these were abandoned short time later due to technical reasons.

From January 1945 all available resources were committed to the construction of the new fuselage nose and the intended weapons installation. On 25/01/1945 Lechfeld again requested the missing installation drawings for the MK 214. Meanwhile, in February the wooden mockup was completed in Oberammergau, and soon afterward this was fitted with a MK 214 (possibly also a mockup). After this a first prototype nose with functioning MK 214 was built. On 23/02/1945 Hauptdienststellenleiter Saur received a brief telex from the Messerschmitt factory:

"Weapon 214 in aircraft (the Me 262 – author's note) fired up to five shots in succession on the stand. Automatic firing starts tomorrow. Grouping very good."

The first forward fuselage from Oberammergau finally arrived in Lechfeld on 11 March 1945. Karl Baur undertook the first flight at the beginning of March. Messerschmitt report No. 56 covered the entire Me 262 program between 01/03 and 31/03/1945 and for the first time named the Me 262 A-1a prototype with the Werknummer 111899 and the MK 214 AV2:

On March 14 logged several flights with a total duration of 4 hours 57 minutes, all dedicated to MK 214 testing. The new nosewheel undercarriage fairing was lost during the first flight. The aircraft had previously logged five flights. A report written on 06/04/1945 also gave information on prior testing of the Me 262 with MK 214: The 47 rounds fired on the test stand by the Me 262 A-1a/U4, WerkNr. 111899 and the 81 in flight were not entirely satisfactory. There had been a number of ammunition belt feed problems as well as issues with the gun's electrical system. Problems with the

Close-up view of the nose section of the Me 262 A-1a/U4 (WerkNr. 111 899) with its imposing MK 214 AV2 cannon, taken in the summer of 1945.

ejection of spent shell casings also had to be resolved as quickly as possible.

Previously, on 05/04/1945, Major Herget had carried out a live gunnery flight and fired six shells at a ground target near Lechfeld. Lindner followed on 07/04/1945, firing two rounds before the gun failed again. One day later Herget fired another five rounds before the MK 214 AV2 jammed again. Nine test rounds were fired on 09/04/1945, but once again the gun's performance was not entirely satisfactory.

The second prototype, WerkNr. 170083, completed at the beginning of April 1945m was fitted with the MK 214 AV3. According to Messerschmitt records, it also still displayed significant shortcomings.

The narrow belt feed, in particular, caused problems and had to be modified as quickly as possible. On 16 April 1945 Herget flew two fighter missions in the first A-1a/U4. There were several problems with the weapon, denying Herget any chance of success.

Work was under way on a third nose weapon installation which, after the problems encountered to date had been overcome, was to be fitted on the future production version of the Me 262 A-1a/U4, the E-1. It was intended that the modern EZ 42 gyro-stabilized gunsight should be installed.

American forces captured the second A-1a/U4, armed with the MK 214 AV3, at Lechfeld ("Wilma Jeanne", later "Happy Hunter II"). A third MK 214 A was also captured there. The second Me 262 A-1a/U4 was lost just short of its destination while being ferried to Cherbourg, from where it was to be shipped to the USA. Messerschmitt test pilot Ludwig Hoffmann parachuted to safety. The fate of the first prototype is not known.

Prototype of the weapons compartment with mountings for the super-heavy cannon in the nose of the Me 262 A-1a/U4.

"Wilma Jeanne" with pilots of Watson's team, also known as "Watson's Whizzers".

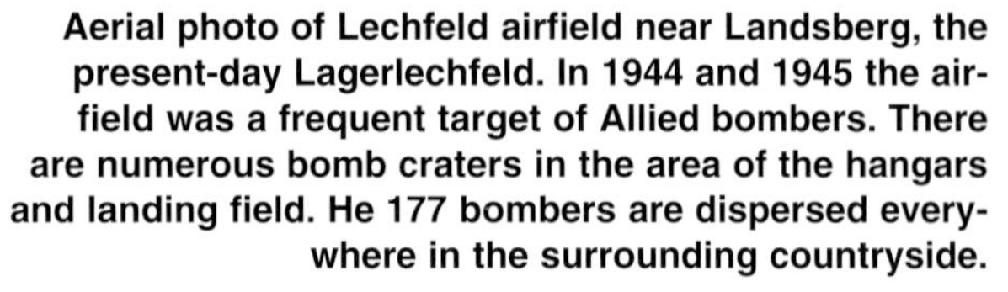
Aerial photo of Lechfeld airfield near Landsberg, the present-day Lagerlechfeld. In 1944 and 1945 the airfield was a frequent target of Allied bombers. There are numerous bomb craters in the area of the hangars and landing field. He 177 bombers are dispersed everywhere in the surrounding countryside.

Prior to leaving Lechfeld "Wilma Jeanne" was renamed "Happy Hunter II". The aircraft never made it to the USA, however, as it was lost en route to Cherbourg.

After testing, one of the captured Me 262s was acquired by American millionaire Howard Hughes. In the 1990s it was offered for sale as a flyable aircraft with its original power plants.

An amateur photo of the REMAG flight line on the Walpersberg near Kahla.

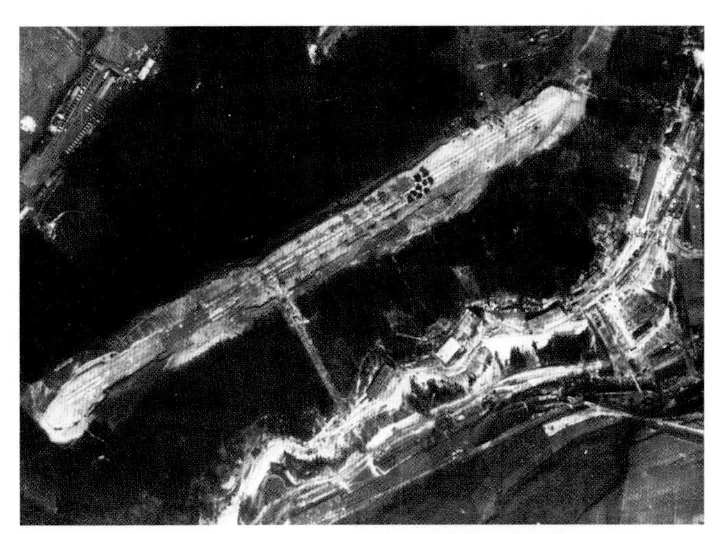

Codename "Lachs" (Salmon)

Me 262 production in the underground facility at Kahla.

Left: reconnaissance photo of the runway on the Walpersberg near Kahla. Note the tracks of the freight lift that was used to transport

Right: Photo of the freight lift track.

Test-flying activity at the Kahla factory airfield on the Walpersberg plateau. The completed aircraft were transported to the top of the hill by means of the inclined freight lift.

Oberstleutnant Heinz Bär, commander of III./EJG 2 in Lechfeld, in front of his Me 262 A-1a "Red 13".

Oberstleutnant Heinz Bär on the wing of his "Red 13". He would be the world's most successful jet fighter pilot when the war ended.

An Me 262 A-1a of III./EJG 2 captured by American troops in Lechfeld at the end of the war.

Right: Fueling operational aircraft of "Kommando Nowotny" at Achmer airfield.

A Kettenkrad tows an operational aircraft of "Kommando Nowotny" to the takeoff line.

Photographs of Me 262 B-1a two-seaters are extremely rare. Here is one such aircraft of KG 76.

Below: The first Me 262 B-1a during flight trials. The two-seat version was built by Blohm & Voss.

These two Me 262s were captured in Lechfeld at the end of April 1944. The aircraft were prototypes of the Me 262 C-1a "Heimatschützer" (see Page 29).

In the spring of 1945 near Regensburg, Me 262 A-1a fighters were repaired and serviced in a former circus tent. This degree of improvisation was due to the prevailing materiel shortages at the end of the war.

The Me 262 V303 (WerkNr. 170303) was used in Lechfeld as a test-bed for development of the "Blitz bomber". The boat-shaped bomb racks just aft of the nosewheel bay door are plainly visible.

"Blitzbomber"

The long-awaited invasion of France began on 6 June 1944. Allied troops landed in Normandy, exploiting their vast material superiority. The fierce fighting was still under way when pilots of 3./ KG 51 began conversion training in Lechfeld near Landsberg on 20 June. The fast "Blitzbombers" were recognizable by the large numbers on the forward fuselage, just below the cockpit. Slow deliveries of bomb racks ("Viking Ship"), the installation of an additional fuel tank, and less major airframe modifications delayed the arrival of the jet-powered fighter-bomber, however.

Senseless directives, such as the one requiring pilots to maintain a minimum altitude of approximately 4000 meters over enemy territory, reduced bombing accuracy. As well, the small number of serviceable "Blitz bombers" limited their efforts to several nuisance raids.

By the time the newly-formed Einsatzkommando (also called Kdo Schenk, Kdo Edelweiss or simply EKdo 51) arrived in Chateaudun on 20 July 1944, St. Lo and Caen in Normandy had already fallen and the port of Cherbourg was encircled by enemy troops.

The supplies needed for the use of jet bombers on forward airfields did not arrive. Allied aircraft attacked the German infrastructure in large numbers and destroyed fuel and spare parts before they could reach the front. And so by the end of July 1944 there were just three or four serviceable Me 262s.

Under enemy pressure, on 12 August the unit had to be withdrawn to Étampes. On 15 August 1944 Kommando Schenk pulled back to Creil and just eight days later to Juvincourt, where on 23 August the unit was bolstered by

Trials were carried out with RI-503 takeoff-assist rockets in an attempt to shorten the takeoff run. These units were indispensible when the aircraft was carrying a full load of 1000 kg of bombs.

A Me 262 A-2a loaded with two SC 250 bombs.

elements of the 3. Staffel of KG "Edelweiss". Only five of the nine Me 262s made it to the airfield, however. There were three crashes and one pilot landed far from his destination. On 28/08/1944 seven Me 262 "Blitz bombers" flew four missions from Juvincourt against Allied troop assembly areas. They mainly used AB 500 bomb dispensers, of which there were only about 20 available at the end of August, filled with SD 10 anti-personnel bombs. At the end of the month the "Blitz bombers" attacked Melun near Paris. One SC 500 landed on the city center, another the area of the railway station, while two AB 500s were dropped on nearby wooded areas suspected of sheltering enemy troops.

The rapid Allied advance forced the unit to fall back to Ath-Chièvres in Belgium on 28/08/1944 and to Volkel-Eindhoven in the Netherlands two days later. One aircraft arrived there with its undercarriage extended. During the transfer flight it had been attacked by several Spitfire fighters and the "Blitz bomber" caught fire. The pilot nevertheless succeeded in landing, in which the aircraft was seriously damaged. The Me 262 was blown up during the retreat. On 4 September 1944 the new base of operations was attacked by the RAF and the Kommando was forced to withdraw to Rheine in Westphalia in the night and fog.

The unit continued operations with its remaining five (originally nine) Me 262s until 13 September 1944. As a rule the targets were about 250 kilometers from the Kommando's base, for example in the Liège area. In September two pilots failed to return from operational sorties. Most of the pilots agreed that several losses had been due to navigational errors and the unaccustomedly high speed of their aircraft. Another complicating factor was the compass' tendency to spin for several minutes after each diving attack. As well, the FuG 16 ZY and FuG 25a radios only functioned properly in about five percent of cases. This hampered the use of Y control or "Tornado" homing.

Beginning the end of September, the jet bombers attacked the Nijmegen bridges and numerous military targets in the area of the 1st Canadian and 2nd British Armies more or less successfully. The "Blitz bombers" subsequently employed SD 250 high-explosive bombs and AB 250 bomb dispensers over the Chièvres, Eindhoven, Nijmegen and Volkel airfields.

By November 1944 KG 51 had achieved a level of accuracy equivalent to a radius of 100 meters. The new personnel reaching the unit had not been trained in dive-bombing attacks, however, and were rarely able to achieve this feat.

On 20 November 1944, I./KG 51 had 28 Me 262s, while the II. Gruppe in Schwäbisch-Hall had just 15 of its authorized complement

Ground crew using an hydraulic trolley to load a bomb onto a Me 262's ventral stores rack.

Operational Me 262 in the winter of 1944-45.

The ETC 504 stores rack, used on later versions of the Fw 190, was also tested on the Me 262.

An SC 500 bomb beneath the "Viking Ship" bomb rack. In the upper left of the photo are the spent shell casing ejector ports for the MK 108 cannon.

A Me 262 A-2a of 1./KG 51 in Rheine. The unit's principal targets were in the Ardennes, northern Germany and in the Bad Kreuznach area.

of 40 "Blitz bombers". III./KG 51 had up to 14 Me 262s, one Ju 88, several Me 410s and a number of Bf 109s and Fw 190s for conversion training. Fourteen flying instructors were responsible for training 116 pilots on the new type. The former III. Gruppe, which had been equipped with Fw 190s, had been renamed 1./ KG 10 in July 1944 and operated the Fw 190 F and G.

The German offensive in the Ardennes began on 16 December 1944, but the Me 262s of the Edelweiss Geschwader did not join the fighting until six days later. Until the end of December the unit's pilots flew numerous low-level attacks against targets in the snow-covered Ardennes. Several Ar 234 B-2s of KG 76 also participated in these operations. On 27 December 1944, for example, eight of the jet bombers joined the Me 262 A-2s in attacks in the Bastogne area.

In mid-December 1944 preparations for "Operation Bodenplatte" entered their decisive phase. Beginning at 07:45 in the morning, the "Blitz bombers" of KG 51 took off to strike enemy forward airfields in the Brussels, Arnhem-Eindhoven and Venlo areas.

There followed a number of low-level attacks in the northwest, before elements of KG 51 were forced to withdraw to Giebelstadt. For a brief time they attacked targets in Alsace, which was finally lost on 18 January, however.

In mid-January the remaining Staffeln flew attacks in the Kleve area and the western Rhineland. The Allied pressure did not let up, however, and the German units were steadily forced back.

On 13 February 1945, therefore, the OKL ordered that "the focus of operations in the ongoing struggle, with bombs and guns by the jets by day and the night close-support units by night, is the frontline area between Nijmegen and Schleiden. Fighter operations, on the other hand, will be scaled back."

The jet bombers flew a large number of missions in the west in a short time. Despite a heavy air raid on Rheine on 24 February, one day later "Blitz bombers" rose to attack Allied targets in the area Jülich, Kalkar and the western Ruhr region.

The first attack by jet bombers against the bridge at Remagen took place on 7 March. These continued for the next two weeks. Soon afterward, on 13 March 1945, the majority of the Me 262s had to be moved back to Leipheim (I./KG 51). From there their operational targets were targets in the southwest of Germany, for example enemy armor that advanced as far as the Pfalz. At least two attacks followed on 19 and 23 March 1945 against Allied advance roads in the Bad Kreuznach area. On 30/03/1945 the "Führer's General Plenipotentiary for Jet Aircraft", General der Waffen-SS Dr. Kammler, ordered that from then on KG 51 was to be attached to IX. Fliegerkorps under General Kammhuber to take part in the defense of the Reich. Kammhuber subsequently ordered 2/3 of the Geschwader's Me 262s transferred to JG 7 and 1/3 to KG 54 (J).

Several days later Göring was convinced to overturn this move. On 4 April 1945 the OKL war diary noted that the transfer of the unit to IX. Fliegerkorps was to be cancelled, especially since Dr. Kammler had ordered the quasi-disbandment of KG 51 without orders, and that the "Blitz bombers" were to continue operating as fighter-bombers as per Hitler's instructions.

By then the Allies had stepped up their attacks on the airfields in the Rheine area. Soon afterwards the last "Blitz bombers" left their bases in Westphalia.

The order to release its aircraft to fighter units at the beginning of April resulted in a drop in KG 51's actual strength in Me 262s. Restoration of the unit's strength therefore began on 10 April 1945. On 21 April the I. Gruppe moved to less-threatened Memmingen, even though its strength was still low. At the same time, 11./KG 51 withdrew into the Nuremberg-Erlangen area. New targets were located in the Würzburg area. Almost simultaneously, a "Blitz bomber" attack took place against the still intact bridge over the Danube near Dillingen. Despite the use of nine jet bombers the bridge escaped damage. On 22 April the bulk of 11./KG 51 was overrun by tanks of the American 7th Army at the Strasskirchen airfield. One day later the few surviving Me 262 A-1s and A-2s were handed over to JV 44, a fighter unit based in Munich-Riem. Five pilots were attached briefly to 1./KG 51 before being transferred to Galland's unit.

Soon afterward the Geschwaderstab (unit headquarters) of KG 51 was disbanded. It was decided to concentrate all of the unit's Me 262s into one operational group and transfer them to Munich-Riem. Two days later, early on the morning of 26 April 1945, Luftflottenkommando 6 (Air Fleet Command 6) ordered all of 1./KG 51's serviceable Me 262s flown to Prague-Ruzyne, where they would come under the operational control of IX. Fliegerkorps (J). There, under the command of Hauptmann Abrahamczik, the elements of the unit supported German forces engaged in the Battle of Berlin and at the end of April carried out low-level attacks against lines of communication in the rear of the 3rd and 4th Soviet Tank Armies. Basically the "Blitz bombers" in Prague served to bolster Gefechtsverband Hogeback. Between 28 April and 5 May 1945 there followed numerous dive- and level bombing attacks against targets in Bohemia-Moravia. After the Czech uprising began, the jet bombers were also called upon to undertake low-level attacks in the Prague area. They also took part in two fighter operations together with several Me 262 A-1a fighters of Gefechtsverband Späte (formerly the Stab and III. Gruppe of Jagdgeschwader 7 "Nowotny").

On 7 May 1945 the remnants of KG 51, now become a Gefechtsverband or battle unit, withdrew to Saaz, present-day Zatec in the Czech Republic. A few hours before the war ended, four pilots flew their Me 262s to Fassberg, Lüneburg and Munich-Riem.

The first use of jet-propelled combat aircraft as bombers and low-level strike aircraft had come to an end.

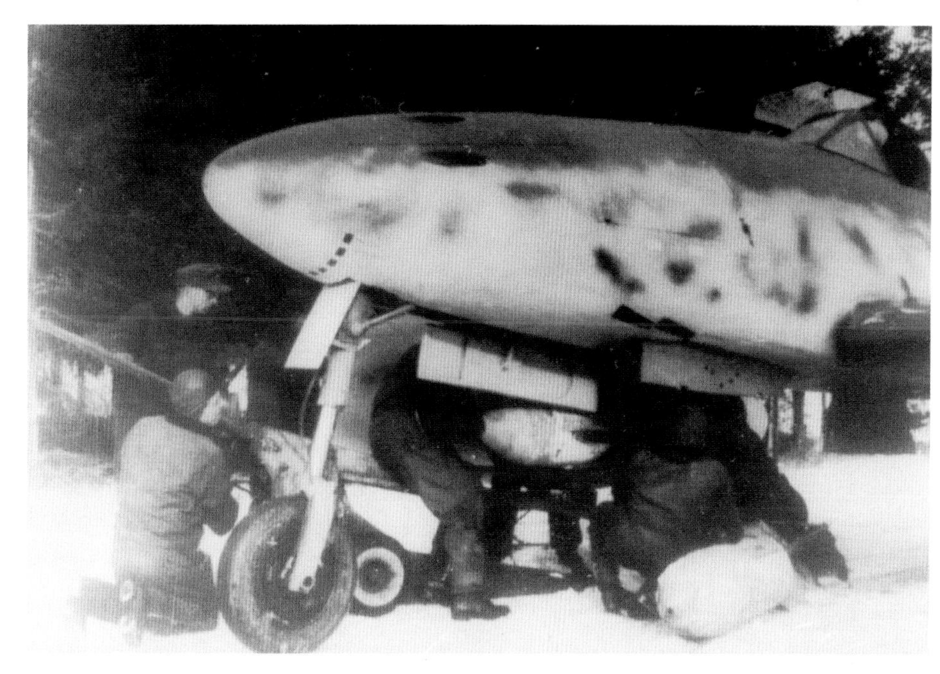

Members of the ground crew practice loading bombs onto a Me 262 in winter conditions.

Parked Me 262 A-2a "Blitz bomber".

The "Viking Ship" bomb racks are plainly visible on the underside of this Me 262.

An operational aircraft of 1./KG 51 in the spring of 1945.

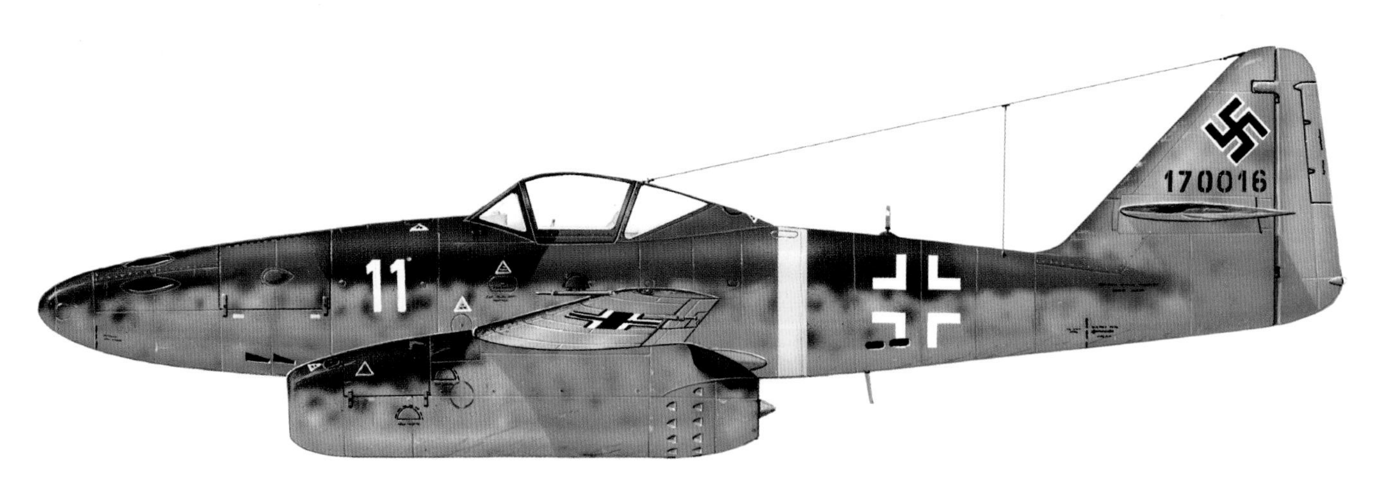

Three-view of a typical operational aircraft finished in the three-color scheme of RLM 74/75/76. As applied at the factory, this was a high-contrast scheme, however it faded quickly in service.

Graphic: Manfred Meyer

The only example of the Me 262 C-2b in Lechfeld.

Prototype of the BMW 003 R in Munich-Allach.

A drawing from the offer description for the Me 262 C-2b.

Me 262 "Heimatschützer II"

The history of few other Me 262s is so well-documented forty years after the war as that of the Heimatschützer II, or Me 262 C-2b.

Despite this, more than five years of extensive research were required to gather information on all of the flights and test stand runs.

What emerged was the picture of a very special Me 262. To believe that it was a project of the very last days of the war would be wrong. The development history of this interceptor in fact began in 1941, during feasibility studies of pulse-jet engines similar to those later used in the V1 program.

On 15/06/1942 an order was issued for 10 prototypes and 100 pre-production examples of the BMW parasite power plant, which was designed for installation on modified BMW 003 turbojet engines.

Precise performance calculations began in the summer of 1943. The resulting data for the Interceptor II was convincing. It appeared that the sought-after aircraft to defend against British bombers had been found.

Peter Kappus, who was involved in the program, described how the aircraft was to have been used: the TLR aircraft (turbojet engine with rocket boosting) would take off when enemy formations approached. Climbing slowly using jet propulsion, they would be vectored until the enemy was sighted. At that point the TLR fighters would be about 2000 meters below the bombers. Only then would the rocket engines be started, after which the Me 262s could reach and attack the enemy machines in a matter of seconds. The decisive advantage compared to the Me 163 was that the TLR fighters could operate on turbojet power only between attacks and thus had greater range and endurance.

So much for the plans. At the time there wasn't even a single prototype with which to test these tactics.

Between October and December 1943 there was much discussion about converting the Me 262 A-1 or A-2 into a TLR interceptor with BMW P 3390 boosted power plants. On 14/01/1944 a brief description of the new Heimatschützer or "Homeland Defender" was issued, followed by a description of the components that would have to be modified.

The first live test run of a BMW 003 R took place in March 1944 and a good 50 more followed. A mockup of the interceptor's airframe was used to assess the effects of the hot exhaust gases on the fuselage and tail section. Fifteen successful test runs were made in Berlin-Spandau in May 1944 alone. Earlier, on 28/04/1944, the project handover for the Me 262 C-2b (the "b" indicated the use of BMW power plants) had been submitted. Actually 15 BMW 003 compound engines were supposed to have been available by then, however technical problems had caused repeated delays. Work on the fuel system, the fuel dump system and the acid-resistant fuel tanks, which had begun in June 1944, was still not completed in the spring of 1945. Work on the engine mounts and cowlings for the BMW 003 R had been completed quickly, on the other hand. The static testing of the Me 262 as a Heimat-

Explosion of the BMW P 3395 rocket unit on 25/01/1945.

schützer had also been carried out with positive results. A performance comparison completed on 23 August confirmed the superiority of the Me 262 C-2b compared to the contemporary Me 163 B-1 in almost all respects.

The next months were filled with additional performance calculations and additional protective measures against leaking Sv-Stoff (sodium nitrate and sulfuric acid).

The actual history of the C-2b interceptor began on 20/12/1944. The airframe selected for conversion (WerkNr. 170074) finally arrived in Lechfeld. Two days later, work to repair discovered airframe damage and modify the electrical circuitry was proceeding apace. At the end of December a newly-manufactured tank ventilation system was installed in the fuselage. In the first days of January the four MK 108 cannon were removed and replaced with ballast. Operational use of the aircraft was thus out of the question. As well, new attachments were installed for the two TLR power plants and the tank ventilation system was again modified.

On 8 January 1945 a maintenance test flight was carried out using pure jet propulsion without the two BMW P 3390 booster engines. The entire aggregate power plant had meanwhile been given the designation BMW 3395 (003R) and the RLM number 109-718.

The engines were checked over after the test flight and problems were found with the starters, resulting in replacement of one of the two Riedel starters. As well, metal shavings were found in the gearbox bearings of the left TLR. The entire power plant was subsequently removed.

Four days later a "water run" was carried out on the starboard rocket unit. One day later the same procedure was carried out on both power plants. The purpose was to check the tightness of the fuel lines without first having to work with dangerous rocket fuels. After a leak was discovered in the pump system, on 14/1/1945 the port engine was removed. Then on 14 January the Riedel starter in the BMW 003 A failed three times.

Four days later the starboard TLR was replaced. Not until 21 January were both aggregates again clear for bench runs. It was found that the Sv-Stoff ventilation line was too small, preventing the required combustion chamber pressure from being achieved. Not until 24 January did both TLR power plants display the necessary benchmarks for a live test run. The aircraft was fueled with Sv- and R-Stoff (rocket fuel components) for the next day.

On 25 January 1945 there was an explosion in the combustion chamber of the starboard engine, completely overturning the testing schedule. The P 3395 compound engine was badly damaged, and the fuel lines were significantly affected. Lechfeld lacked the facilities needed to repair the engine, so a replacement was ordered from BMW.

Below: The special (rocket) fuel scoured the paint from the aft fuselage of WerkNr. 170004.

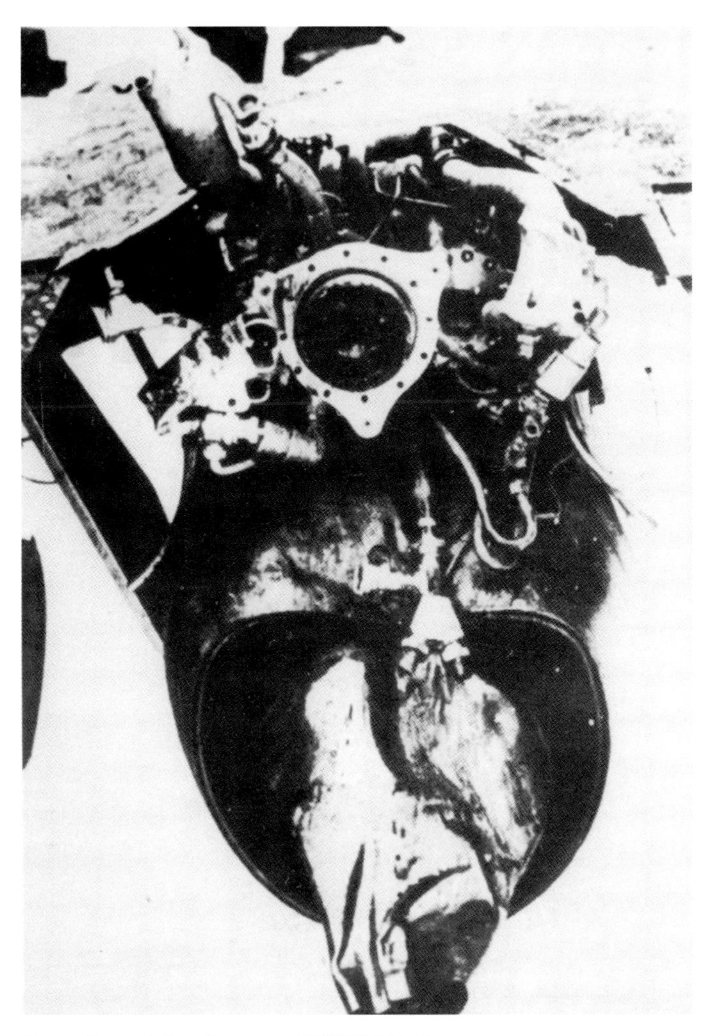

Rear view of the destroyed BMW 003 R.

Preparations for the first flight with TLR propulsion on 26/03/1945.

On 1 February tests with the still intact TLR continued. Bench runs were carried out, three very short ones and two of 70 seconds. Further tests had to wait, however, as the Sv-Stoff had by then eaten through the seals. Four days later a stationary run was carried out, however the pressure retaining valves and seals leaked and the pressure buildup distended the fuel tank.

The new fuel tank system ordered on 11 February arrived in Lechfeld on 14/02/1945. Two days later another test run was made, during which the only failure affected the Riedel starter.

Live runs on 18 February also failed to produce satisfactory results. There were fuel pressure problems, the ignition system failed to work and the fuel pumps failed again. On 21 February, therefore, the port TLR was removed. After the source of the problems could not be found, the next day a new engine was installed.

By 23 March there were ten failures of the Riedel starters, until finally a successful bench test was completed. Although the TLR power plants were now operational, the aircraft could not be flown due to a shortage of B-4 jet fuel.

Three days later, on 26 March 1945, company test pilot Karl Baur climbed into the Heimatschützer II, the Me 262 C-2b, and took off on the first successful TLR flight. The two compound engines were switched on for 40 seconds in the climb and accelerated the aircraft considerably.

After a thorough examination of the TLR power plants, the Heimatschützer made its second and ultimately final test flight in Lechfeld on 29/03/1945.

A switch problem prevented the rocket boosters from being used. The cause of the problem could not be ascertained as there was no more fuel for the jet engines.

For this reason both TLR aggregates were removed and the aircraft was set aside. On 27/04/1945 the Me 262 C-2b was captured by the Americans and was probably scrapped in the autumn of 1945.

Engine test run in March 1945.

Me 262 Variants

Unlike other prototypes that never reached the production stage, the variants illustrate don this page were brought to operational maturity and saw live action.

The Me 262 A-1a/U3 reconnaissance variant seen in the photo on the left was employed operationally by a number of NAGs (tactical reconnaissance Gruppen).

The Me 262 A-1a V056 (WerkNr. 170056) with the radar installation (FuG 218 Neptun) ultimately led to the Me 262 B-1a/U1 night-fighter version, which was used operationally by "Kommando Welter".

Above: Me 262 A-1a/U3 in operational use.

Left: Me 262 A-1a/U3 WerkNr. 170111 after a take-off crash caused by engine failure. Schwäbisch-Hall, 1 January 1945. The cannon armament was removed and replaced by Rb 50/30 cameras.

Me 262 A-1a V056 with a FuG 218 "Neptun" air intercept radar installed for test purposes. Lessons learned from this trials machine were incorporated into the later Me 262 B-1a/U1 night-fighter.

Captured Me 262 B-1a/U1 at Wright-Patterson Air Force Base. The aircraft had previously served with "Kommando Welter".

Me 262 V10 during trials with a towed payload. The arrangement was difficult to control and development was soon cancelled.

Gerd Lindner takes off with a towed 3payload on 22/10/1944. The arrangement was called the "towbar method".

Lindner in flight near Lechfeld with towed payload.

This Me 262 A-1a fitted with two launch tubes for WGr. 42 210-mm rockets belonged to III./JG 7 "Nowotny". The officer on the left is Rudi Sinner.

An aircraft of the Third Gruppe of JG 7, also equipped with two rocket launching tubes. The low-velocity 210-mm rockets were quickly replaced by R4M rockets, which were also unguided.

Operational aircraft of "Jagdverband 44", formed in March 1945, photographed near Munich. The unit was made up of the fighter arm's most experienced aces and tried to change Germany's fortunes in the air. Hampered by shortages of fuel and spare parts, however, it accomplished nothing extraordinary.

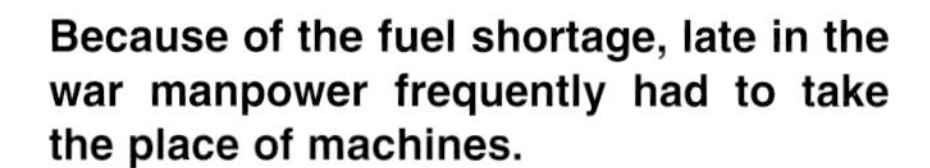

Because of the fuel shortage, late in the war manpower frequently had to take the place of machines.

Me 262 ("White F", WerkNr. 111685) captured by Allied troops on the Munich-Salzburg autobahn. The aircraft was attached to JV 44.

Cockpit of the Me 262 A-1a.

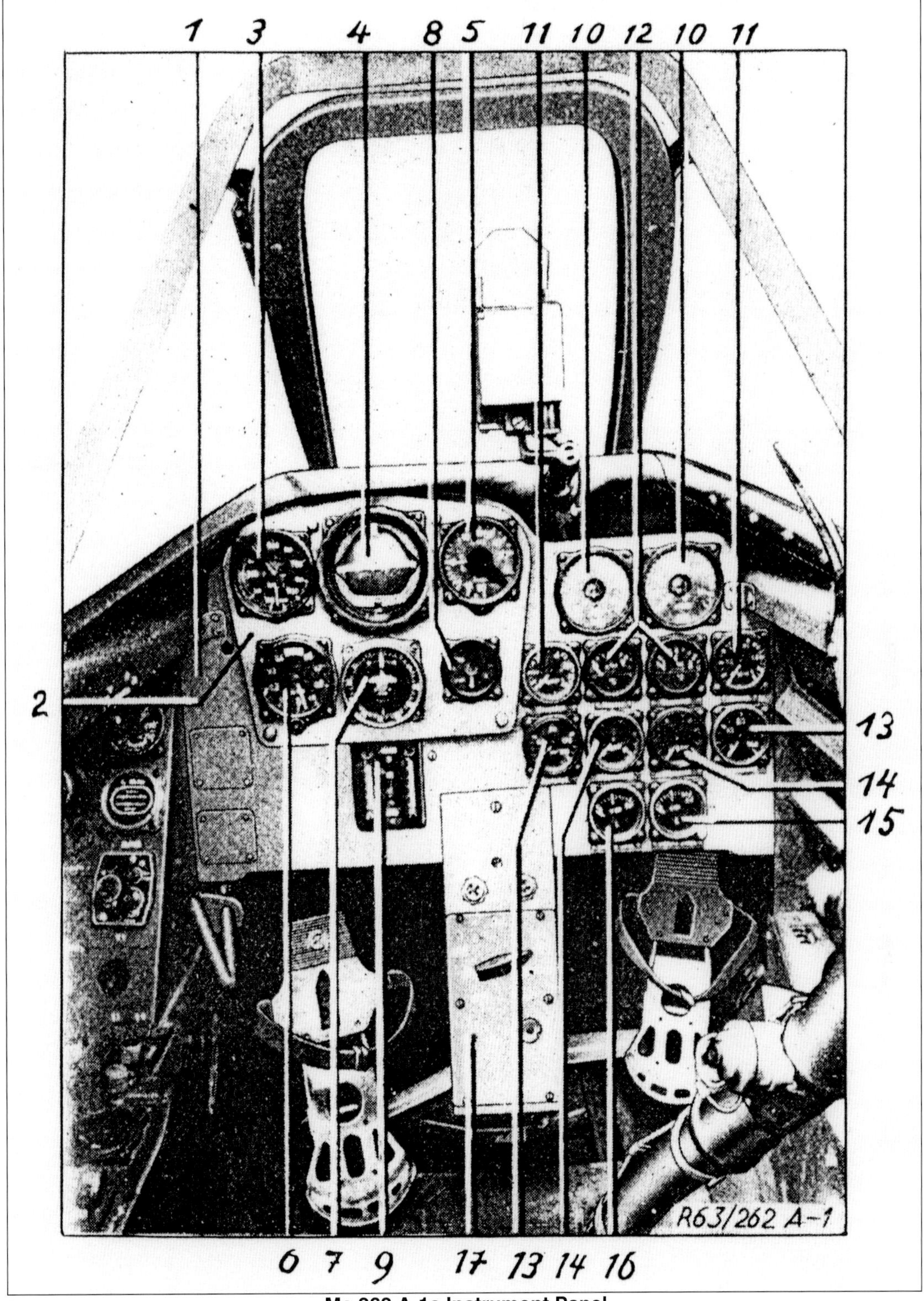

Me 262 A-1a Instrument Panel

1. Instrument Panel
2. Blind flying panel
3. Airspeed indicator
4. Combined turn-and-bank indicator and artificial horizon
5. Vertical speed indicator
6. Altimeter
7. Master compass indicator
8. AFN 2 homing indicator
9. SZKK 2 ammunition counter
10. RPM Indicator
11. Gas pressure meter
12. Gas temperature indicator
13. Pressure gauge (injection pressure)
14. Pressure gauge (oil pressure)
15. Fuel contents gauge (forward)
16. Fuel contents gauge (rear)
17. ZSK 244 A (only installed in fighter-bombers)

Wrecked Me 262s in the final assembly hall at the Messerschmitt factory in Obertraubling.

The End...

The collapse of Germany also meant the end of flying for the jet fighter units. Along many main supply routes lay wrecked Me 262s that had been crash-landed, abandoned or shot down. Some planes were documented by Allied soldiers with their cameras.

These photos form important historical documents today.

Below: Me 262 A-2a, WerkNr. 501221 "Yellow 3". The aircraft was shot down by American ground forces near Dresden-Klotsche in April 1945. The Defense of the Reich fuselage band identifies the aircraft as belonging to III./JG 7.

Right: Unfortunately we have been unable to positively identify where this photo was taken. Some sources state that it was a final assembly line along the Leipheim autobahn. Others describe the location as a dispersed final assembly line in Obertraubling.
Mitte rechts)

Center right: A natural metal Me 262 A-2a (WerkNr. 111711) discovered by the Salzburg autobahn. The aircraft was later taken to the USA and tested.

Below: American troops found Me 262 A-1a WerkNr. 500490 near the Austrian border next to the Munich to Salzburg autobahn. When found the aircraft was still intact, however it was very quickly stripped by American souvenir hunters.

Below right: Discovered in Austria, this Me 262 A-1a "White 22" wears an extremely dark camouflage scheme. It is not known where this photo was taken.

This Me 262 A-1a of JG 7 ended up in England. The aircraft is now in the Royal Air Force Museum in Hendon.

The Survivors

A considerable number of Me 262s were captured and survived the war. After extensive ground or flight testing they found their way into various museums around the world. The aircraft illustrated on this page are a representative selection of the aircraft preserved in the USA. Other examples may be found in England, Australia and South Africa.

This Me 262 A-1a – converted from a U4 – is now on display at the US Air Force Museum in Dayton, Ohio.

One of the finest restorations of a Me 262 was carried out by the team at the National Air and Space Museum in Washington. The Me 262 A-1a, WerkNr. 500491, was previously operated by 11./JG 7.

Above: This spurious camouflage finish was applied to a captured Me 262 in the USA. The subsequent fate of the aircraft is not known.

Right: Difficult to believe that this rudimentary display object – part of a display of captured enemy aircraft at Wright Patterson Air Base – would one day become the legendary "Yellow 7" seen at the bottom of Page 40.

Event though the camouflage scheme is not entirely authentic, the technical condition of this Me 262 on display at the Planes of Fame Museum in Chino, California is very good. There is serious interest in making the machine flyable again.

The Messerschmitt Me 262 B-1a is prepared for its ferry flight to Cherbourg by Watson's Whizzers.

The Story Continues...

Taken to the USA, the Me 262 B-1a was later reborn as "Black 35".

The Me 262 B-1a once parked at Naval Air Station Willow Grove later provided data for the planned production of five Me 262 copies under the direction of Herbert Tischler in his Texas Airplane Factory.

As the plans needed to build this small series of Me 262 replicas were incomplete, they came up with the idea of restoring an existing example and – if had previously been disassembled – create the missing drawings in the process. The operation was a success and after successful restoration Herbert Tischler was able to return the Me 262 B-1a to its original owner as "Black 35". Left behind was an invaluable treasure trove of data, which made it possible to proceed with the small batch of Me 262 A-1a replicas.

First stage in the final assembly of "Black 35".

Another photo of the final assembly at the Texas Airplane Factory owned by Herbert Tischler.

Right: Standing on its own feet with the power plants already in place.

Below: The completed "Black 35" prior to its return to Naval Air Station Willow Grove.

Rollout of the first Me 262 at Paine Field. The aircraft have interchangeable fuselage spines and can be flown as single- or two-seaters.

The Swallow Is Brought to Life...

In 1998 the Texas Airplane Factory encountered financial difficulties and was unable to proceed with the Me 262 project.

Bob Hammer, a well-known aircraft designer, who was in the process of retiring, was asked if he might continue the project. When he agreed, the project was moved to Paine Field on the west coast. There the first aircraft, "White 1", was completed and extensively test flown.

The photos presented here were taken during various stages of development and the aircraft's maiden flight.

The aircraft are disassembled for painting. In the background is D-IMTT destined for the Messerschmitt Foundation.die für die Messerschmitt-Stiftung vorgesehene D-IMTT zu sehen.

Above: A series of wonderful archive photos were taken before the aircraft was readied for its maiden flight.

Left: Back in the hangar for tests, tests and more tests.

Below: The big moment. With Wolfgang Czala at the controls, "White 1" lifts off from the runway. It was the first flight by a Me 262 in more than 50 years.

The maiden flight was followed by a whole series of test flights. One ended in a serious undercarriage failure which set the program back by months.

Right: Two former enemies meet again at Paine Field.

There has to be a little show.

Above: With Wolfgang Schirdewahn at the controls, the aircraft returns from a successful training flight.

The Legend Lives On.....

On 15/07/2005, with Wolfgang Czaia at the controls, the Me 262 A-1a for the Messerschmitt Foundation took off from Paine Field on its maiden flight, which proved uneventful. After several weeks of flight testing and approval procedures, the aircraft was delivered to the Messerschmitt Foundation in Manching.

Since then the aircraft has taken part in many public flying displays, causing the hearts of onlookers to beat faster.

The aircraft is regularly flown by test pilots Wolfgang Schirdewahn and Horst Philipp.

After more than twenty years, the idea, the execution and the permanent refinement of the "Me 262 Replica Project" is an unqualified success.

Below: The aircraft taxis to the Messerschmitt Foundation's hangar.

The unsurpassed sight of a Me 262 "Schwalbe" in flight.

Wolfgang Schirdewahn on final approach to Manching airport.

The camouflaged Me 262 blends in well with the gray sky.

The Me 262 turns onto final. The unsurpassed sleekness of the aircraft's lines is obvious from this angle.

The Me 262 performing at an air show.

Me 262 A-1a of KG 51 (J) in Neuburg an der Donau. The pilots of KG 51 converted to the Me 262 straight from Junkers Ju 88s after just three flights. Because of lack of experience and low serviceability levels they achieved little success in the fighter role.

ı contrast to the photos on the facing page, conditions on German airfields at the end of the war were no longer satisfactory. Intact unways and taxiways were scarce, making jet operations extremely dangerous.

US $14.99
51499
9 780764 340482
ISBN: 978-0-7643-4048-2
150005